REASONING SKILLS IN MATHS

Years 3 & 4

Talk it, solve it

by Jennie Pennant

with Claire King and Jacky Walters

We would like to thank the following schools in Bracknell Forest for trialling these activities:

Ascot Heath Infant	Meadow Vale Primary
Ascot Heath CE Junior	New Scotland Hill Primary
Binfield CE Primary	Owlsmoor Primary
Birch Hill Primary	The Pines Primary
Broadmoor Primary	Sandy Lane Primary and Nursery
College Town Infant	St Joseph's Catholic Primary
College Town Junior	St Margaret Clitherow Catholic Primary
Cranbourne Primary	St Michael's CE Primary
Crown Wood Primary	St Michael's Easthampstead CE VA Primary
Crowthorne CE Primary	Uplands Primary
Fox Hill Primary	Warfield CE Primary
Great Hollands Primary	Whitegrove Primary
Harmans Water Primary	Wildridings Primary
Holly Spring Infant and Nursery	Winkfield St Mary's CE Primary
Holly Spring Junior	Wooden Hill Primary and Nursery

We would also like to thank the BEAM Development Group.

BEAM Education

BEAM Education is a specialist mathematics education publisher, dedicated to promoting the teaching and learning of mathematics as interesting, challenging and enjoyable. Their materials cover teaching and learning needs from the age of 3 to 14 and they offer consultancy and training.

BEAM is an acknowledged expert in the field of mathematics education.

BEAM Education
Nelson Thornes
Delta Place, 27 Bath Road
Cheltenham GL53 7TH
Telephone 01242 267287
Fax 01242 253695
Email cservices@nelsonthornes.com
Orders orders@nelsonthornes.com
www.beam.co.uk

Published by BEAM Education

© BEAM Education 2005

All rights reserved.

The photocopiable activities in this book may be reproduced by individual schools without permission from the publisher.

Reprinted in 2011.

ISBN 978 1 9031 4277 6
British Library Cataloguing-in-Publication Data
Data available
Edited by Raewyn Glynn
Designed by Malena Wilson-Max
Printed by Multivista Global Ltd

Contents

Preface

Our primary teachers in Bracknell Forest have developed these collaborative activities to give children aged 5–11 more opportunities to engage in meaningful mathematical discussion. They have ensured that talking and working together are central to the activities, and have trialled them extensively in our primary schools to make sure that they work. Teachers reported that the children engaged enthusiastically in the dialogue, and in the thinking that ensued.

The main objective is to get children talking about mathematics. The context for this is logical thinking and reasoning, because there is scope here for debating more closely mathematical definitions, properties and patterns. Our teachers observed that the children did question each other, and wanted justifications, before they reached consensus.

We, in Bracknell Forest, hope that these activities will support and stimulate colleagues in promoting mathematical talk in the classroom, and that children's enjoyment and understanding of mathematics will be enhanced as a result. Many thanks to the teachers, and children, who took part in the development of these materials.

A. Fletcher

Allison Fletcher

Assistant Director
Education, Children's Services and Libraries
Bracknell Forest

Introduction

Recent initiatives in mathematics teaching emphasise the importance of mathematical discussion within problem-solving, and within mathematical learning generally. Children learn maths by doing it and talking about it, hence language is integral to securing mathematical learning. The activities in this book require children to focus on speaking and listening, and on reasoning, as they interpret clues and identify which items to eliminate.

When we talk, we engage in dialogue with others, and we receive feedback from them. We want children to be involved in mathematical dialogue to help them explore, investigate, challenge, evaluate and actively construct mathematical meaning. When children work together in a small group, they can articulate their thinking, listen to one another and support each other's learning in a safe situation. Organising problem-solving in small groups increases the potential for developing the skills of speaking, listening and working together.

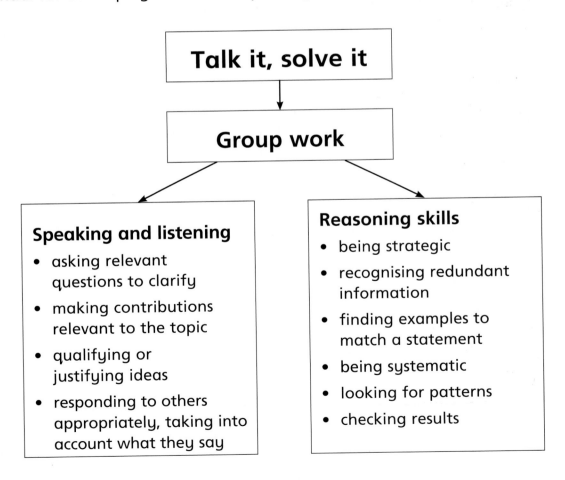

Talk it, solve it

↓

Group work

Speaking and listening
- asking relevant questions to clarify
- making contributions relevant to the topic
- qualifying or justifying ideas
- responding to others appropriately, taking into account what they say

Reasoning skills
- being strategic
- recognising redundant information
- finding examples to match a statement
- being systematic
- looking for patterns
- checking results

The 'Talk it, solve it' activities

In these activities, children identify an unknown item (number, shape, amount, and so on) by means of clues, or questions and answers.

Each unit contains:

- a 'Solve it' sheet giving a collection of eight items, one of which is the unknown item.

- a 'Talk it' sheet with eight clues. These clues give enough information (and more) to identify this unknown item.

- an 'Ask it' sheet with a set of questions. Children choose their own item for others to identify, and the rest of the group ask the questions to discover the unknown item. Some of the questions are deliberately left open by including a blank space where children can insert their own number, shape, amount or other item.

The first unit consists of the 'Solve it' sheet, three different 'Talk it' clue sheets, and an 'Ask it' sheet. This unit will enable you to introduce the activity to the whole class, and to give the class confidence in using clues logically and effectively.

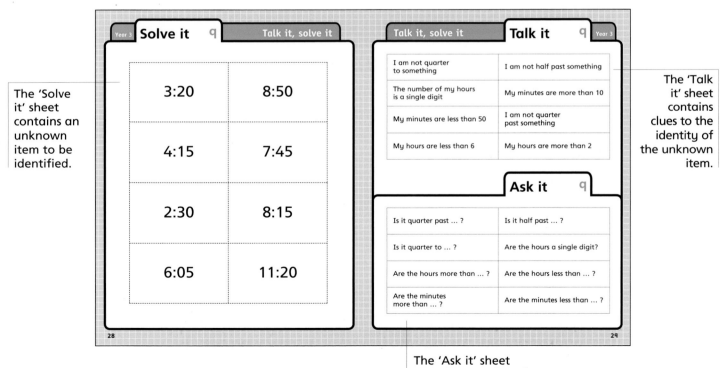

The 'Solve it' sheet contains an unknown item to be identified.

The 'Talk it' sheet contains clues to the identity of the unknown item.

The 'Ask it' sheet contains questions for use in group work.

Introducing the activity to the whole class

The 'Talk it' and 'Solve it' sheets

Display the 'Solve it' sheet on the whiteboard or overhead projector. Alternatively, if these are not available, give each pair of children a copy of the 'Solve it' sheet. Look at the numbers (or other items) shown and invite the children to identify some of the properties of the numbers. Support their observations by asking questions such as "Do you think that number is in the 5 times table? I wonder if there are any other numbers in that table?"

Display the 'Talk it' sheet, look at the clues shown and talk about them briefly, or read out each statement in turn. You may choose to use the opportunity to check that children understand the mathematical content of the clues. Discuss with the class how to use these statements to eliminate items on the 'Solve it' sheet.

Choose one statement from the 'Talk it' sheet and cross out any eliminated items on the 'Solve it' sheet as they are identified. Continue like this until there is one item left. Go through any remaining clues as a check. If any statements don't fit, challenge the class to redress this or rethink.

The answers to the 'Solve it' sheets can be found on page 36 (for Year 3) and on page 64 (for Year 4).

The 'Ask it' sheet

Display the 'Ask it' sheet and read through the questions. Tell the children you are going to think of an item) on the 'Solve it' sheet. They must ask you questions to discover the item you have secretly chosen. Discuss with the children which items to eliminate from the 'Solve it' sheet for each question.

When they have worked out your item, invite a pair of children to choose a secret item from the 'Solve it' sheet. Other pairs decide which questions to ask to find out the secret item. Talk with the children about the best questions to ask.

Working in pairs or small groups

Each group will need the 'Solve it' sheet to share between them, and the 'Talk it' sheet cut into strips. Ask the children to take it in turns to choose a 'Talk it' strip, read it to the group and together decide which item to eliminate on the 'Solve it' sheet. The children should keep taking turns until they have only one item left on the 'Solve it' sheet. Extend the activity by asking the group to establish if all the

'Talk it' statements are equally important. Ask if some statements eliminate more possibilities than others. Ask how few statements they can use to solve the problem. These questions could be used as a basis for a plenary discussion.

Meeting diverse needs

Children who need additional support

The activity can be adapted for children who need additional support by asking them to take a 'Talk it' strip and match it to one of the items on the 'Solve it' sheet. The activity then becomes one of identifying properties of a number, shape or other item.

For those who can manage a further challenge, put the cut 'Talk it' strips in a pile and invite children to pick the top one from the pile. Talk this through to decide which items it could refer to. Dealing with one clue at a time is simpler for inexperienced children than choosing a strip from the complete range.

More able learners

Suggest that children explore which of the 'Talk it' statements is the most informative and eliminates most items. When they have identified the unknown item on the 'Solve it' sheet, they should try again and find how many different routes there are through to that solution, including the shortest and the longest.

Invite them to invent their own questions as well as using those on the 'Ask it' sheet. Pairs can invent their own 'Solve it' sheet and a set of clues for other pairs to solve.

Those learning English as an additional language

Adapt the activity by helping children to organise themselves. Provide two sheets of paper labelled 'maybe' and 'no'. Cut the 'Solve it' sheet into eight pieces and put them all on the 'maybe' paper. Then, as the 'Talk it' clues are read out, children can move the 'Solve it' items on to the 'no' paper.

Setting up the activity in this way provides a lot of helpful repetitious questions and statements.

Using the CD-ROM

The CD-ROM contains all the activity sets in the book. You can either project individual pages onto a whiteboard from your computer or print them out onto acetate sheets and use them on an overhead projector. You can also print the sheets out from the CD rather than making photocopies.

Year 3

Unit 1

Properties of number

Unit 1 consists of the 'Solve it' sheet, three different 'Talk it' clue sheets and an 'Ask it' sheet. This unit will enable you to introduce the activity to the whole class, and to give the class confidence in using clues logically and effectively.

467	284
163	263
405	87
273	227

Talk it 1a

I am a three-digit number	I am less than 420
My tens digit is even	My tens digit is less than 8
My hundreds digit is even	My units digit is not 5
None of my digits are the same	I am less than 270

Talk it 1b

I am less than four lots of a hundred	I am more than 260
My hundreds digit is in the 2 times table	I am greater than a century
My hundreds and my tens digit are different	My tens digit is more than 7
All my digits are even	My ones digit is less than 5

Talk it 1c

I am 3 away from a multiple of 10	Add a hundred to me and the answer is greater than 200
My digits add up to a number less than 12	All my digits are different
I am more than 150	I am less than 200
My ones digit is less than 5	If you count from 100 in tens you will not say my name

Ask it 1

Is it less than 280?	Is it less than ... ?
Is it more than ... ?	Is the tens digit even?
Is the tens digit less than ... ?	Is the hundreds digit 2?
Is the ones digit ... ?	Do its digits add up to a number less than ... ?

46	73
19	17
27	110
55	37

Talk it 2

I am not a multiple of 5	The sum of my digits is 10
I am a two-digit number	Both my digits are odd
My ones digit is less than 8	I am not a multiple of 2
Add 3 to my number and the answer is a multiple of 10	My tens digit is less than my ones digit

Ask it 2

Is it a multiple of 5?	Does it end in … ?
Is the sum of its digits 10?	Is it a two-digit number?
Does it have any even digits?	Is it more than … ?
Is it less than … ?	Is the tens digit more than … ?

38	88
q8	48
q0	118
qq	58

Talk it 3

I am not a multiple of 10	Add 2 to me to get a multiple of 10
I am even	I am less than 100
My tens digit is odd	My ones digit is even
I am not 80 add 10	My tens digit is more than 5

Ask it 3

Is it a multiple of 2?	Is it a multiple of 10?
Is it more than … ?	Is the tens digit more than … ?
Is it a two-digit number?	Is it an odd number?
Is one of the digits odd?	Is the tens digit even?

36	24
120	49
32	48
52	46

Talk it 4

Round me to the nearest 10 and you get 50	Add 100 to me and the answer is less than 150
My tens digit is less than my ones digit	If you divide me by 10 you get a remainder
My tens digit is even	I have two digits
My ones digit is double my tens digit	If you count back in twos from 60 you will say my name

Ask it 4

If it is rounded to the nearest 10, is the answer 50?	If it is divided by 10, is there a remainder?
Does it have two digits?	If I count in twos from … will I say the number?
Is the ones digit even?	Is the tens digit even?
Is the ones digit more than the tens digit?	Is the ones digit double the tens digit?

Solve it   5

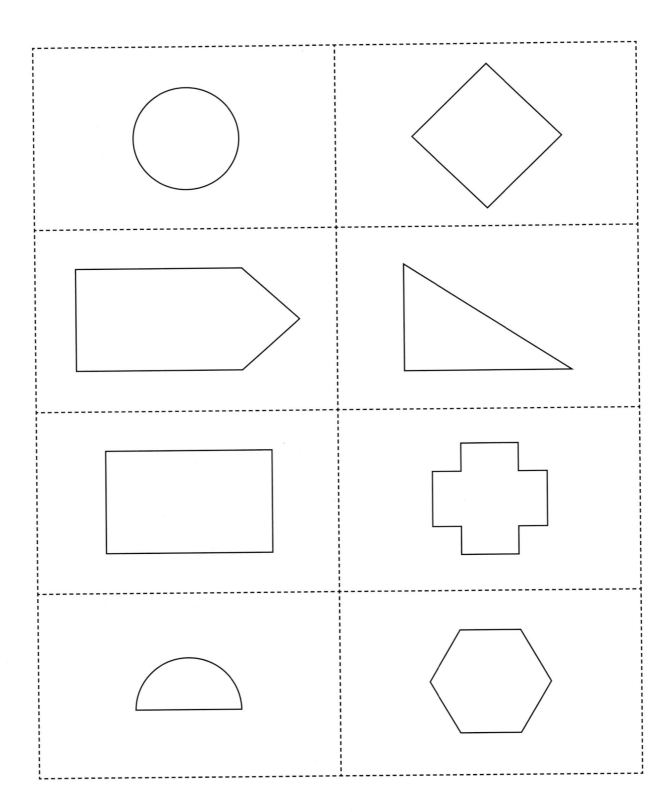

All my sides are straight	All my sides are equal in length
I have more sides than a square	I have no curved sides
I am not a triangle	I have more than three corners
None of my angles are right angles	I have fewer than eight sides

Ask it 5

Has it got any curves?	Are any of its angles right angles?
Are all of its angles right angles?	Are all its sides the same length?
Has it got four corners?	Has it got four sides?
Has it more than ... sides?	Has it got fewer than ... sides?

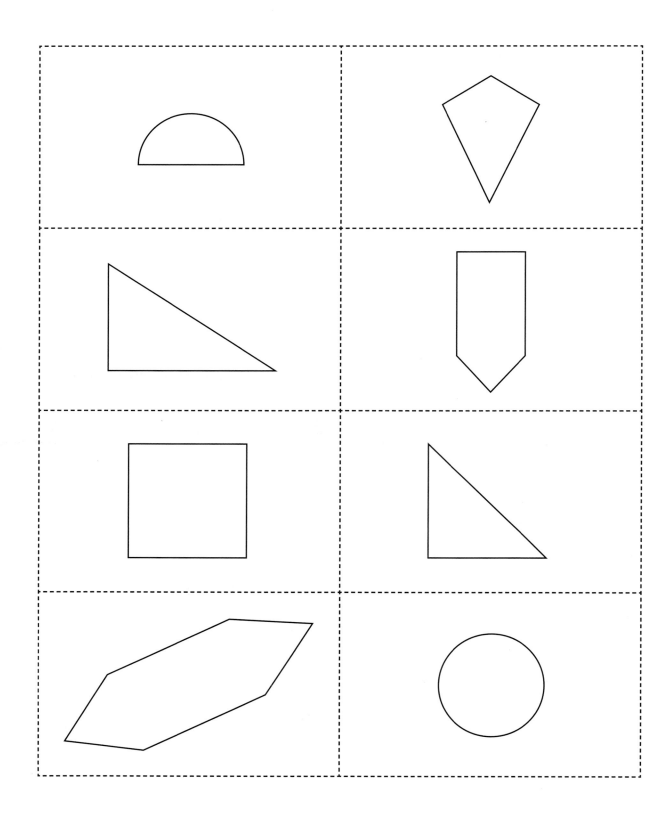

Talk it   6

I am not a hexagon	I am not a semi-circle
I have at least one straight side	I am not a triangle
I do not have 6 sides	I have exactly one line of symmetry
I have more than 3 sides	I have no right angles

Ask it   6

Is it a hexagon?	Has it got any curves?
Has it got more sides than a triangle does?	Has it got exactly one line of symmetry?
Has it got any right angles?	Are its sides all equal in length?
Has it got more than … corners?	Has it got fewer than … sides?

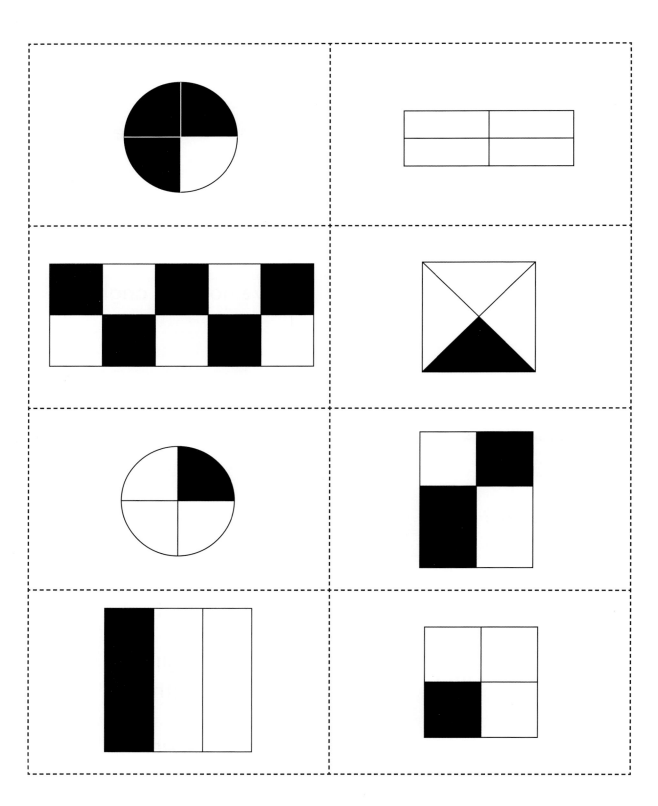

I am not divided into tenths	I am not divided into thirds
I am divided into four parts	I am not divided using diagonals
I am divided into equal parts	Some of me is shaded
Less than $\frac{1}{2}$ of me is shaded	I have no curves

Ask it   7

Is it divided into equal parts?	Is it divided into quarters?
Is it divided into thirds?	Has it got any curves?
Is it divided using diagonals?	Is $\frac{1}{4}$ of it shaded?
Is more than … of it shaded?	Is less than … of it shaded?

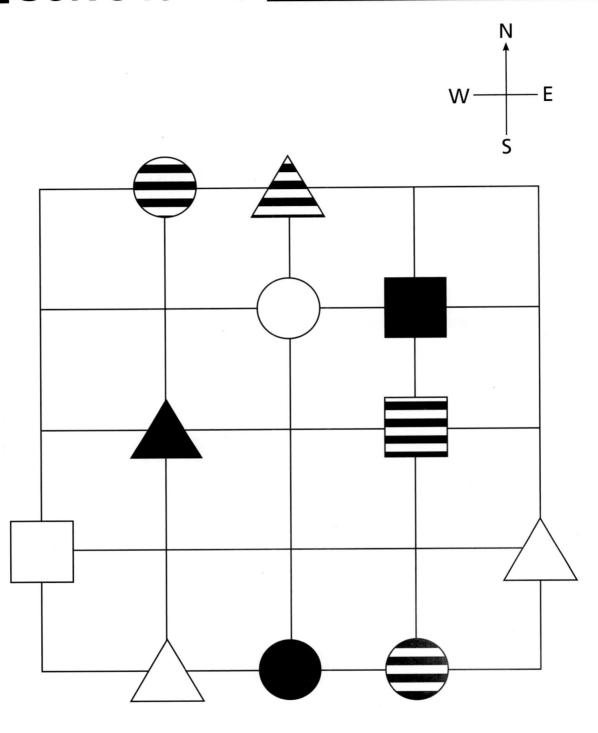

Talk it　8　Year 3

I have another shape on the same horizontal line as me	There is a circle on the same vertical line as me
I am either black or striped	I am west of a stripy shape
There is a stripy shape north of me	There is a white shape south of me
I am not white	I am black

Ask it　8

Is there a … shape on the same horizontal line?	Is there a … shape on the same vertical line?
Is it north of a … shape?	Is it south of a … shape?
Is it the shape that is furthest east on the grid?	Is it the shape that is furthest west on the grid?
Is there a … north of it?	Is there a … west of it?

3:20	8:50
4:15	7:45
2:30	8:15
6:05	11:20

I am not quarter to something	I am not half past something
The number of my hours is a single digit	My minutes are more than 10
My minutes are less than 50	I am not quarter past something
My hours are less than 6	My hours are more than 2

Ask it   q

Is it quarter past … ?	Is it half past … ?
Is it quarter to … ?	Are the hours a single digit?
Are the hours more than … ?	Are the hours less than … ?
Are the minutes more than … ?	Are the minutes less than … ?

Solve it 10

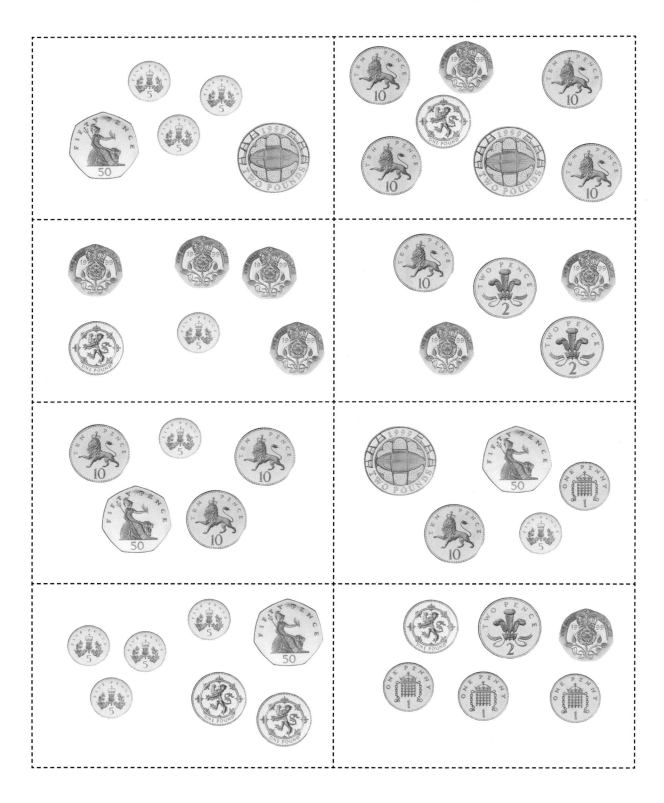

Some of my coins are worth the same amount as each other	I have more than £2.50
I don't have any 50p coins	Five of my coins are each worth less than £1
My total value is worth more than two 50p coins	Two of my coins are worth more than 50p each
I have more than £1	I have a £2 coin

Ask it   10

Are any of the coins worth the same amount as each other?	Are any of the coins worth more than 50p?
Are any of the coins worth less than … ?	Are there any 50p coins?
Are there any 5p coins?	Is there more than one … coin?
Is the total value more than … ?	Is the total value less than … ?

400 g

650 g

600 g

700 g

750 g

1500 g

800 g

900 g

I weigh less than 1 kilogram	You could balance me using just 100 g weights
Double my weight is more than 1000 g	I do not weigh the same as seven 100 g weights
I weigh more than half a kilogram	If you put me on a balance against a 900 g weight, I would rise up
Put a 500 g and a 100 g weight together, and they weigh less than me	I weigh less than 1000 grams

Ask it 11

Does it weigh more than 1 kilogram?	Does it weigh less than half a kilogram?
Does it weigh more than … grams?	Does it weigh less than … grams?
Could I balance it using just 100 g weights?	Does it weigh the same as … 100 g weights?
If it was on a balance against a … g weight, would it rise up?	Is its weight a multiple of 100 g?

48 cm

65 cm

82 cm

55 cm

110 cm

64 cm

73 cm

66 cm

Talk it 12

I measure less than seven lots of 10 centimetres	I measure more than half a metre
I am longer than two 30 cm rulers	A metre stick is longer than me
I am an even number of centimetres	I measure less than 66 cm
I am not as long as a metre stick	If you add 20 cm to me, the answer is less than a metre

Ask it 12

Is it longer than a metre?	Is it shorter than ½ a metre?
Is it less than … lots of 10 centimetres?	Is it more than … lots of 10 centimetres?
Is it longer than a metre stick?	Is it an even number of centimetres?
Does it measure less than … cm?	Does it measure more than … cm?

Answers Year 3

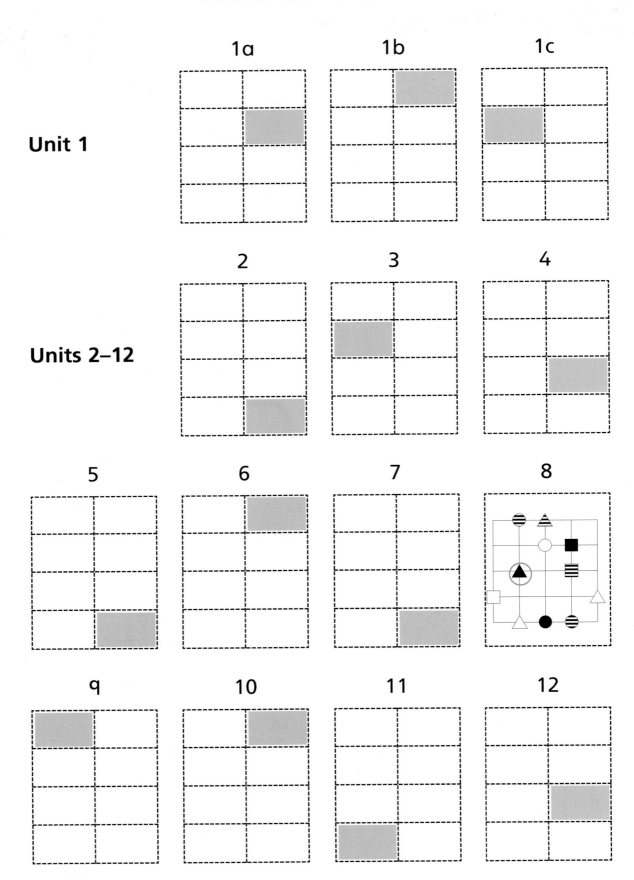

Unit 1

1a   1b   1c

Units 2–12

2   3   4

5   6   7   8

q   10   11   12

Year 4

Unit 1
Properties of number

Unit 1 consists of the 'Solve it' sheet, three different 'Talk it' clue sheets and an 'Ask it' sheet. This unit will enable you to introduce the activity to the whole class, and to give the class confidence in using clues logically and effectively.

16	31
69	72
76	73
12	92

Talk it   1a

I am less than 90	My digits add up to an odd number
I am even	I am more than 70
My tens digit is odd	If you rounded me to the nearest ten you would get 70
My tens digit is greater than my ones digit	Count up or down in fours from 60 and you get to me

Talk it   1b

I am less than 9 lots of 10	I am even
The sum of my digits is more than 5	The sum of my digits is less than 10
The sum of my digits is not 10	Double me and the answer is less than 100
I am in the 4 times table	Count back from 80 in twos and you'll get to me

I am less than half of 140	I'm more than double 6
Double me is more than 50	I am odd
My digits total a number less than 10	Count in fours and you won't land on me
Both my digits are less than 7	Both my digits are odd

Ask it 1

Is it less than … ?	Is it more than … ?
Is it even?	Is the sum of the digits more than … ?
Do the digits add up to an odd number?	Is the tens digit odd?
Is the tens digit greater than the ones digit?	Is it in the … times table?

Solve it   2

119	139
135	155
99	149
124	129

I am not even	My digits are all odd
I'm less than double 70	I am more than 100
My digits are all different	I am more than half of 220
I am one less than a multiple of 10	Count on in threes from 100 and you will get to me

Ask it   2

If I count on in fives from 100, will I reach it?	Are all its digits odd?
Is double this number less than … ?	Is it more than … ?
Are all its digits different?	Is it odd?
Does it end in 9?	Is it less than double 80?

12	32
50	20
22	44
24	qq

Talk it 3 Year 4

I am a multiple of 4	I am not a multiple of 5
I am not a multiple of 10	I am more than five sixes
I am even	The sum of my digits is even
Double me is more than 80	I am not a multiple of 3

Ask it 3

Is it a multiple of 4?	Is it a multiple of 5?
Is it a multiple of 3?	Is it a multiple of 10?
Is it more than … sixes?	Is it even?
Is the sum of its digits even?	Is double the number more than … ?

Solve it 4

2595	515
2515	2015
3335	2535
2620	2540

Talk it 4

I am a four-digit number	I am more than 2500
I am an odd number	I am less than 3000
My tens digit is not 1	My hundreds digit is 5
The sum of my digits is odd	My tens digit is less than 6

Ask it 4

Is it a four-digit number?	Is the sum of its digits more than … ?
Is the sum of its digits even?	Is it an odd number?
Is it less than … ?	Is its tens digit … ?
Is its hundreds digit … ?	Is its thousands digit … ?

$$\frac{4}{8}$$

$$\frac{3}{8}$$

$$\frac{1}{4}$$

$$\frac{2}{8}$$

$$\frac{1}{2}$$

$$\frac{5}{8}$$

$$\frac{1}{8}$$

$$\frac{3}{4}$$

Talk it 5

I am less than 1	I am worth more than one eighth
I am worth more than one quarter	If you read me out, you won't say the word 'three'
I am not five eighths	I am not three quarters
You use a 4 when you write me down	I am worth less than five eighths

Ask it 5

Is one of the digits a 4?	Is it worth a number of quarters?
Is it worth one half?	Is it worth more than a half?
Is it worth … ?	Is one of the digits a … ?
If you read me out, will you say the word 'three'?	If you read me out, will you say the word 'four'?

Solve it 6

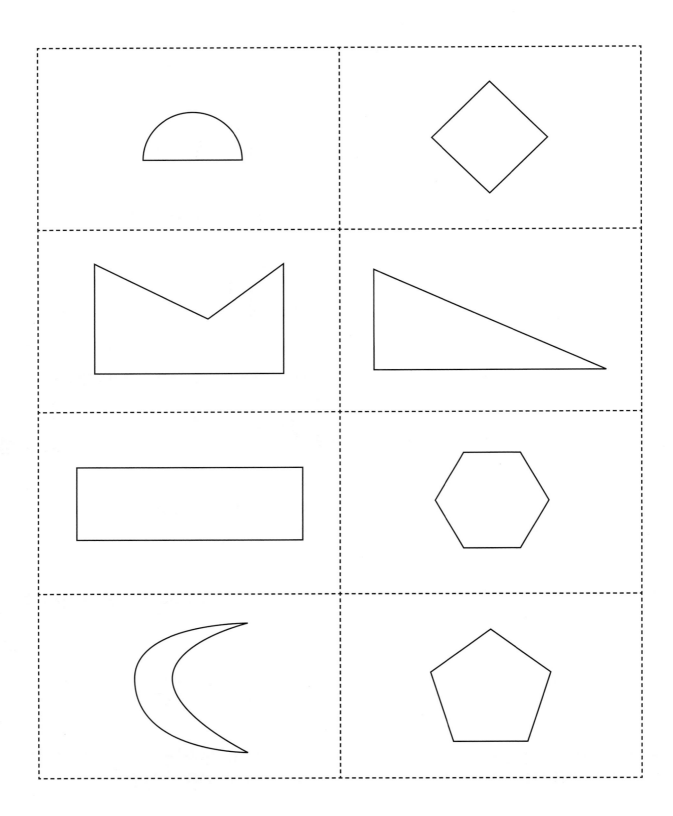

Talk it 6

I have no curves	I have more than two sides
I have at least one right angle	I have fewer than six vertices
I am not a quadrilateral	I have exactly one pair of parallel sides
My sides are not all the same length	I am not a regular polygon

Ask it 6

Is it a regular polygon?	Has it got more than ... sides?
Are all its sides the same length?	Has it got at least one right angle?
Has it got fewer than ... vertices?	Has it got one or more curved sides?
Has it got one or more pairs of parallel sides?	Has it got one or more straight sides?

Solve it 7

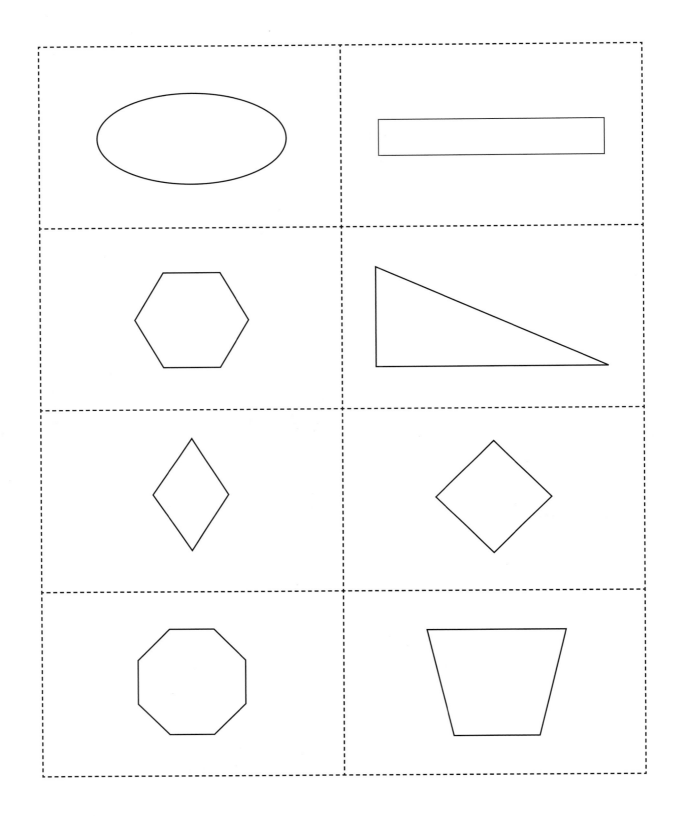

I am not a regular polygon	I have more than two sides
I have no right angles	I have fewer than six vertices
I have exactly one line of symmetry	I have a pair of parallel sides
I have no curved sides	My sides are not all the same length

Ask it 7

Is it a regular polygon?	Are all its sides the same length?
Has it got any right angles?	Has it got fewer than ... vertices?
Has it got more than ... sides?	Is it curved?
Has it got one or more pairs of parallel sides?	Has it got ... lines of symmetry?

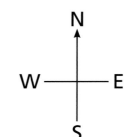

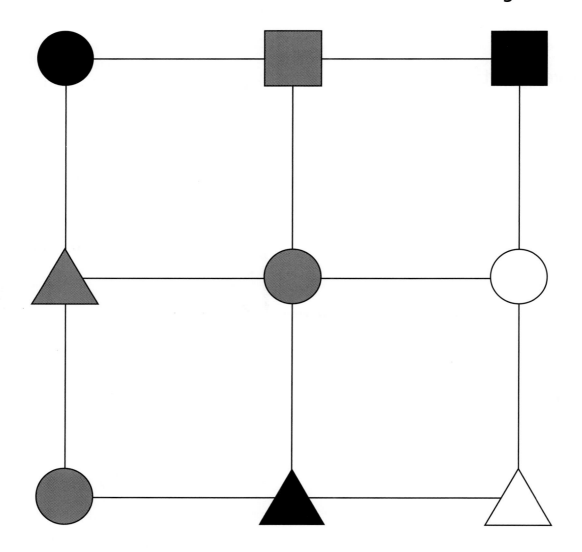

Talk it 8

I am not a square	I am S of a circle
There is a circle N of me	I am N of a grey shape
I am SW of a square	There is no shape to the W of me
I am NW of a shape	I am S of a black shape

Ask it 8

Is it W of a white shape?	Is it SW of a white shape?
Is it NE of a .... ?	Is it N of a ... ?
Is there a shape to the ... of it?	Is there a shape to the ... of it?
Is its colour ... ?	Is it a ... shape?

28th August 2004

1st February 2003

29th March 2006

29th December 2001

5th March 2000

3rd October 1999

3rd February 2001

5th November 2003

Talk it   q   Year 4

I am not in the 20th century	I am not in a leap year
There is an 'r' in my month	My month has 31 days
My month always has the same number of days	My month is not near the end of the year
I am in the last week of my month	I am not in the second half of the year

Ask it   q

Is it in the 20th century?	Is it in a leap year?
Is there an 'r' in the month?	Has the month … days in it?
Does the month always have the same number of days?	Is the month in the first half of the year?
Is the date in the first week of the month?	Is the date in the last week of the month?

Solve it   10

£2.45	325p
£2.81	383p
513p	£3.69
357p	£3.46

Talk it 10

I am worth less than a five pound note	I am worth less than £3.80
I am not worth the same as 346p	I am worth more than seven fifty pences
I have a pound sign	I am worth more than half of five pounds
If you changed me to pennies, there would be an odd number of them	I am worth less than 500p

Ask it 10

Is it worth more than £3?	Is it worth less than £4?
Is it worth more than … ?	Is it worth less than … ?
Is it worth the same as … ?	Is one of its digits a … ?
Has it got a pound sign?	Is there a pence sign?

1.2 kg	1 kg
2000 g	200 g
3000 g	3500 g
2 kg	300 g

Talk it 11

I am between 500 g and 2.5 kg	I am a measurement of weight
I am not equal to 3 kg	I am less than 3500 g
I am heavier than 1 kg	I am heavier than 1000 g
You could balance me with just two weights	My unit is grams

Ask it 11

Is it a whole number of kilograms?	Is the unit grams?
Is it more than 1000 g?	Is it more than … ?
Is it less than 500 g?	Is it less than … ?
Would it balance … 1 kg weights?	Is it equal to a whole number of kilograms?

70 cm	135 cm
1.2 m	90 cm
40 cm	84 m
200 cm	6 cm

Talk it 12

I am longer than half a metre	I am shorter than 2 m
I am longer than a metre rule	I am less than 1 km
I am shorter than 130 cm	I am between 50 and 150 cm
Double me and you will have more than a metre	Halve me and you will have more than 50 centimetres

Ask it 12

Is it written as a number of centimetres?	Is it longer than a metre?
Is it longer than … ?	Is it shorter than 100 cm?
Is it shorter than … ?	Is it between 100 and 500 cm?
Is it between … and … ?	If the length is doubled, will it be more than a metre?

Answers Year 4

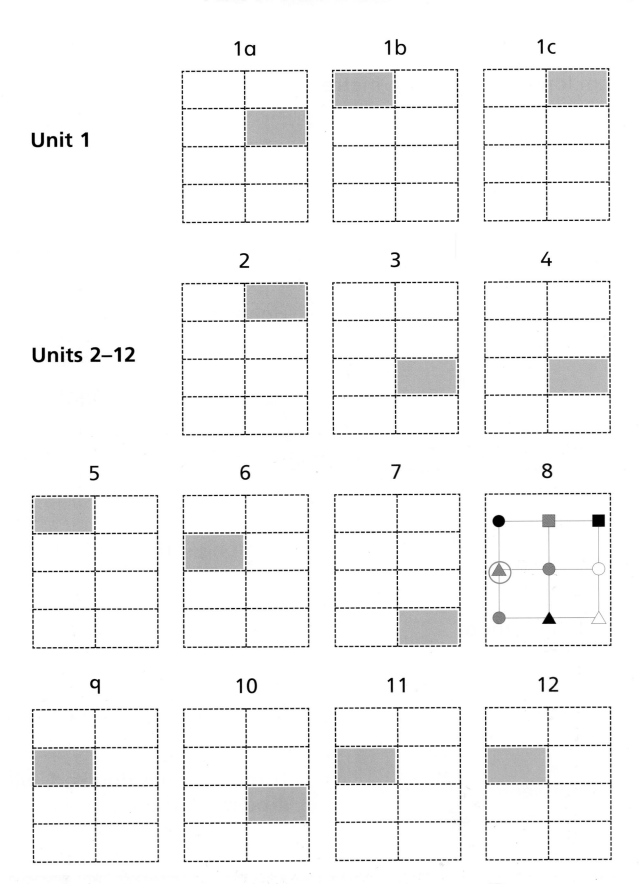

Unit 1

1a   1b   1c

Units 2–12

2   3   4

5   6   7   8

9   10   11   12